CONTENTS

INTRODUCTION TO DIGITAL AUDIO

When Thomas Edison patented the phonograph cylinder in 1878, he never could've imagined what listening to music in the 21st century would be like. Gone are the days when audiophiles needed to collect piles of records, tapes, and CDs to hear their favorite tunes. Now, hundreds of thousands of songs, albums, and recordings are available online for you to listen to at

The digital realm has greatly increased the saturation of music in our everyday lives. This phone can store just as much, if not more, music as the stacks of vinyl in the picture on the next page.

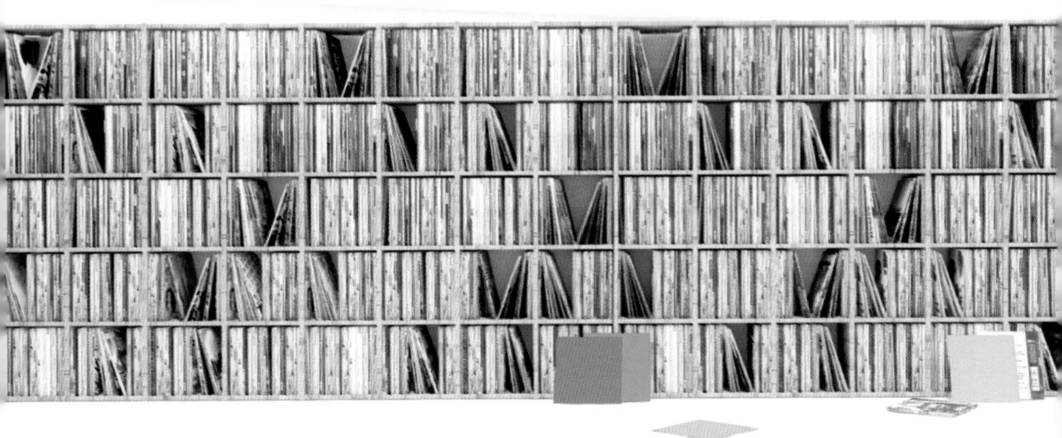

In recent decades the technology that allows us to digitize musical recordings has greatly reduced the amount of space that is required to store music. Vinyl has traditionally been a burden for music lovers because of the massive amounts of storage space they require in homes, the degradation of the recording's playback over time, and their easily damaged, fragile format.

anytime—and they can be easily stored with just a few simple steps. New technologies and ideas come along every day, and companies like Amazon and Apple have recognized the immense potential of our ever-changing digital world.

Smart devices like Amazon's Echo make collecting and storing music a fun, easy experience. And streaming services like Pandora and Spotify have completely changed the way we listen to music. In this booklet, we'll explore some of today's most useful and entertaining choices for listening to your favorite recordings. Whether you want to hear a Frank Sinatra recording from 1958 or listen to a reading of one of the latest *New York Times* Best Sellers, there are some great options to choose from!

WHAT IS AMAZON ECHO?

If you watch television very often, no doubt you've seen the commercials that feature people talking to "Alexa." In the ads, people going about their normal days ask "Alexa" to set alarms, provide weather reports, and play music. Alexa's friendly voice responds accordingly, conveniently delivering information and entertainment to her human counterparts.

This nifty (and seemingly sentient) device is Amazon Echo, a "smart speaker" that connects wirelessly to your internet. The Echo features a voice-controlled "personal assistant"—Alexa, named for Amazon-owned web traffic data company Alexa Internet—which continually listens to speech until it hears the "wake word," *Alexa*. The user can then request information about the weather or news headlines, create shopping or to-do lists, or play music.

Several different Echo devices are available, from the small and relatively affordable Echo Dot—a simple disc that can play music and control smart home devices—to the more elaborate Echo Show—a tablet with a 7-inch screen that can also play video. All Echo devices require an Amazon account before they can be used. A regular Amazon account is free, or you can pay about $100 a year for Amazon Prime. A Prime membership gives you access to streaming video and Amazon Prime Music, so depending on what you use your Echo for, it may be worth the extra cost.

SETTING UP ALEXA

Before you can start listening to music on your Echo, you should make sure your "personal assistant," Alexa, is set up. To do this, you'll need to download the free Amazon Alexa app. This is available for iPhones, Android phones, or your personal computer.

Place your Echo in a central location, like on a kitchen counter, in the living room, or on a bedroom nightstand, and make sure it's plugged in. Now, sign in to the Alexa app, then click ***Settings*** and then ***Set Up a New Device***. Choose your Echo device from the list provided.

The light ring on the Echo should be orange while it's in setup mode. If you don't see orange, press and hold the **ACTION** button—the one with the single dot in the middle—for five seconds. You should hear Alexa say, "Now in setup mode." There should now be guided instructions in the Alexa app that show you how to connect your Echo to a Wi-Fi network. Once you're finished, Alexa will say "Your Echo is ready." The Alexa app will also show you a video to help you learn the basics of your Echo.

Now you're ready to learn about some of the great music options the Echo has to offer!

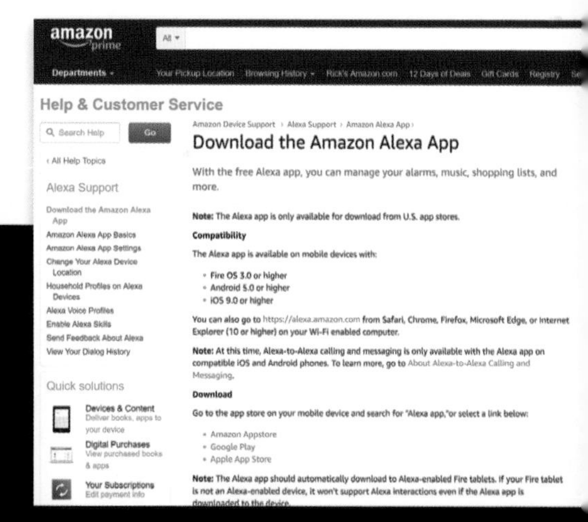

Amazon's Device Support page provides multiple links to locations where you can download the Alexa application for both your smartphone and Wi-Fi enabled computer.

STREAMING MUSIC WITH ECHO

There are many ways to listen to music on your Echo device. The smart speaker comes ready to use with music services including Spotify, Pandora, and iHeartRadio. When you want to listen to a song, podcast, or radio station, open the Alexa app—the same one you used to set up your Echo device. Tap on the three horizontal lines in the upper left corner, and then select ***Music, Video, & Books.***

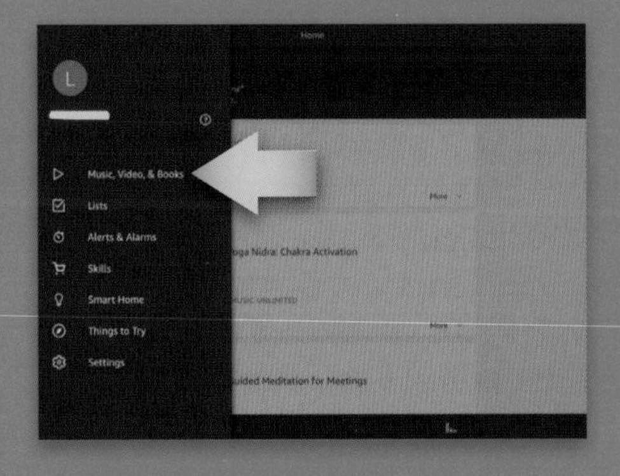

Select the music service you want to use. (It should be noted that in order to use Spotify on your Echo, you'll need a Spotify Premium account, which costs about $10 a month.) If you've never used the service with your Echo before, the Alexa app will ask you to link your Amazon account to the music service. Click on ***Link Your Account Now*** and follow the instructions, then return to the Alexa app. If you'd like, click on the ***Choose Default Music Services*** button at the bottom of the page and select your preferred service.

Now you can scroll down to see the various selections of music offered by the service. When you see something you'd like to listen to, you can either tap it on your device, or simply ask Alexa to play it: "Alexa, play the song [song title]." You can also ask Alexa to adjust the volume, stop playing a song, or tell you who the artist or composer is.

Beginning in 1994, Amazon developed as an online bookstore and has emerged as the world's largest online retailer in 2017.

UPLOADING YOUR OWN MUSIC TO ECHO

Another option for streaming music from the Echo is to use Amazon Music to upload your personal music from iTunes, Google Play, or elsewhere. To upload music, first download and install the Amazon Music app, which is available on the Amazon.com website.

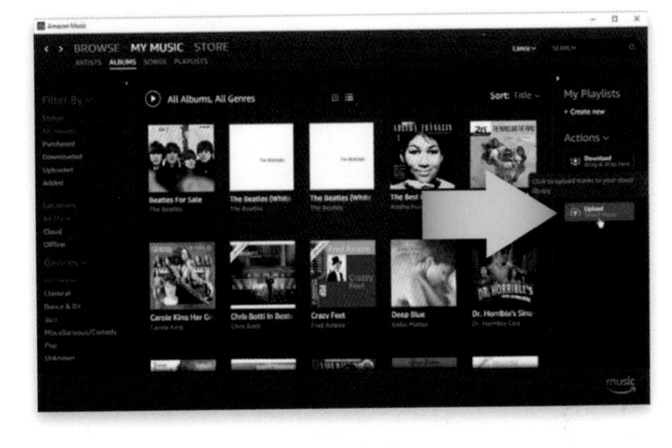

Once you have the Amazon Music app, open it and click on ***My Music*** at the top of the screen. Then click the button on the right-hand side that says ***Upload Select Music***. You can upload specific songs by clicking ***Select Files***, or upload an entire album by clicking ***Select Folder***. Click on the songs or album in your computer's music library that you want to upload, and then click ***OK***. When the music has successfully uploaded, you will receive an "Upload completed" message. You can now ask Alexa to play any of the music you uploaded.

You can only upload 250 songs for free; but if you're an Amazon Prime member, you also have access to Prime Music, which allows you to stream around two million songs. You can also go to the Amazon Music settings page and upgrade your storage plan under ***Music Storage***. For $25 a year, you can upload 250,000 songs and play them all on your Echo.

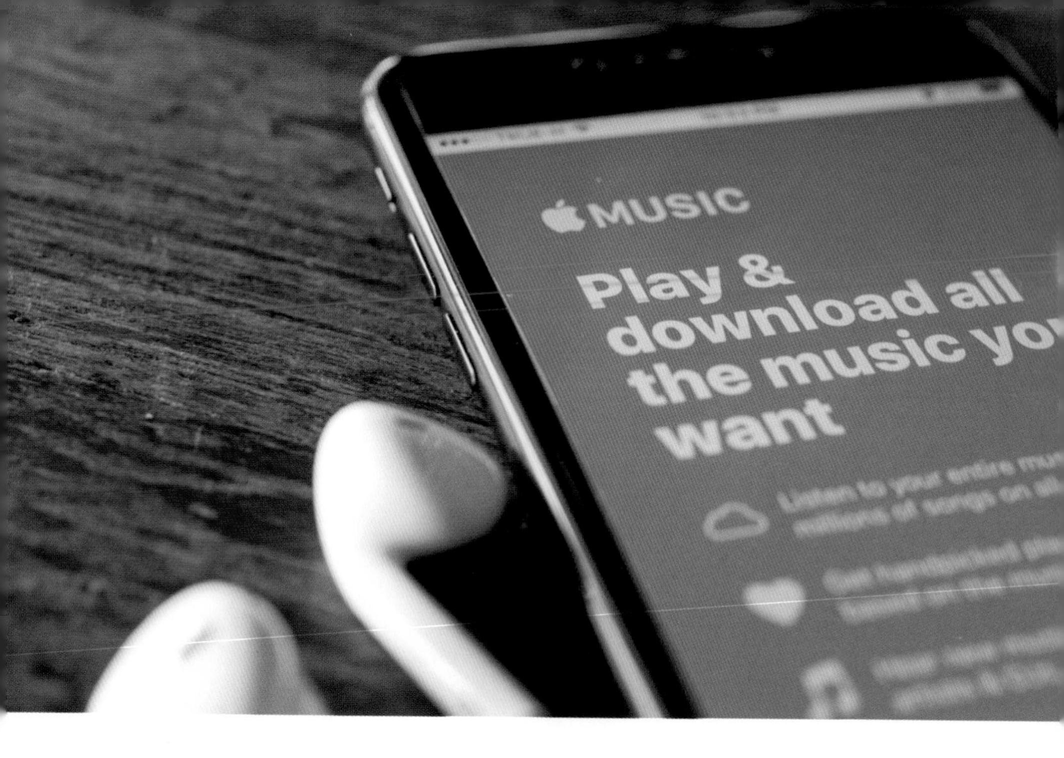

WHAT IS APPLE MUSIC?

Apple Music is a music and video streaming service courtesy of technology giant Apple Inc. The service allows users to select music or playlists that they can stream to devices on demand. It also features a 24-hour radio station called Beats 1 that is broadcast to more than 100 countries worldwide. New subscribers are given a free three-month trial period, and after that the cost is $9.99 a month for single subscribers or $14.99 a month for a family.

MUSIC
ALL THE WAYS YOU LOVE MUSIC. ALL IN ONE PLACE.
Unlimited access to over 30 million songs, a reimagined live radio station, and a personal connection to the artists you love.
START 3-MONTH FREE TRIAL
GO TO MY MUSIC

Apple Music gives you access to 30 million songs from all kinds of genres—so there's something for everyone! And you can download up to 100,000 songs to your own personal library on your Apple devices, or to your PC or Android phone. The service is extremely popular, with more than 36 million subscribers in February of 2018.

SIGNING UP FOR APPLE MUSIC

All new iPhone and iPod models come with the Music app preinstalled. All you need to do is create your account.

If you have an iPhone, iPad, or Apple computer chances are you already have iTunes and access to Apple Music. As long as you're using iOS 8.4 or later on your iPhone or iPad and OS 10.7 and beyond for your computer, the service is preinstalled on your device. To sign up for the service on your iPhone or iPad, tap the app labeled ***Music***. A screen will pop up that welcomes you to Apple Music, and from here you can tap ***Start 3-Month Free Trial***. This will begin the registration process.

You can choose either the individual plan or the family plan, which supports up to six members of your family. (You won't be charged for these plans during the free trial period.) Next, sign in to the iTunes store using your Apple ID and password, and confirm that you want to sign up for Apple Music. You can also sign up on your Mac computer, by launching iTunes and selecting ***Join Apple Music***.

The iTunes app on your computer will give you access to the Apple Music streaming service. It also provides you with a platform to organize all the music files you already own.

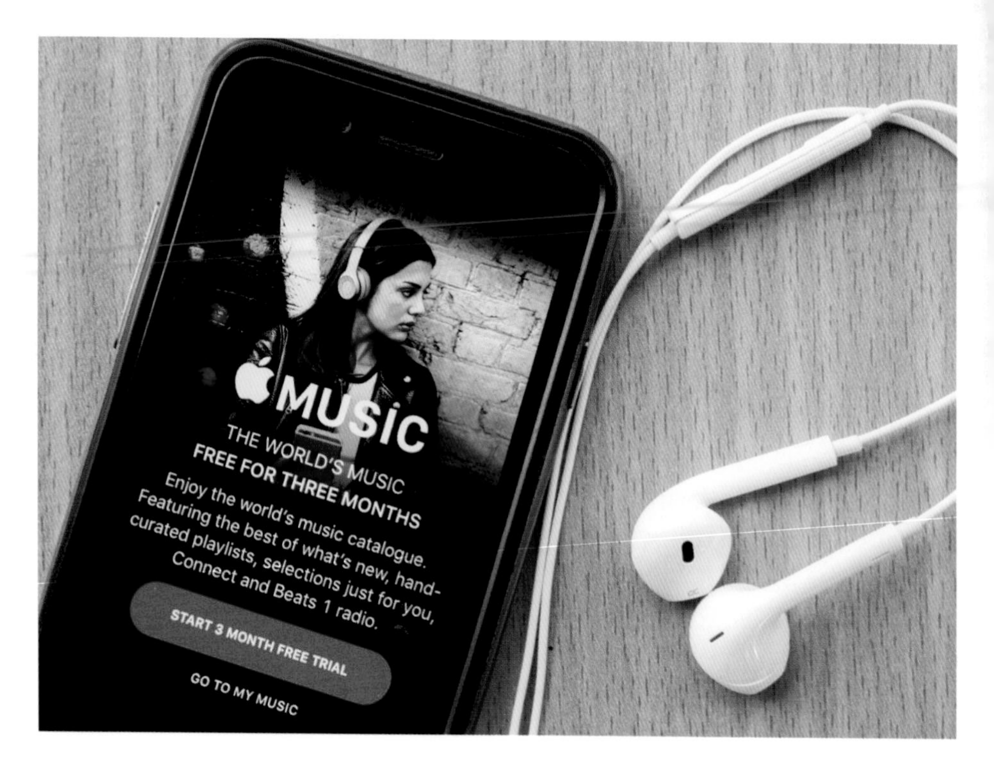

NAVIGATING APPLE MUSIC

When you launch the Apple Music app, you'll see five tabs: ***Library***, ***For You***, ***Browse***, ***Radio***, and ***Search***. The ***Library*** tab is where you'll find your playlists. You can also search for artists, albums, and songs, and see your recently added music.

In the ***For You*** tab, you'll find playlists that the Apple Music app creates based on your music selections, so they're created with your personal tastes in mind. And the ***Browse*** tab allows you to search new and up-and-coming music from many different genres. Use these tabs to sample new music and songs—you may find something to add to your collection!

The ***Radio*** tab is just that: Apple's live Beats 1 radio station, which is broadcast in more than 100 countries. You can also listen to the radio on-demand, in case you miss an interview you really want to hear. And

It is very easy to keep your music organized with Apple Music. Songs should automatically categorize themselves under specific artist and album names. You can also create your own playlists to provide the soundtrack for any situation.

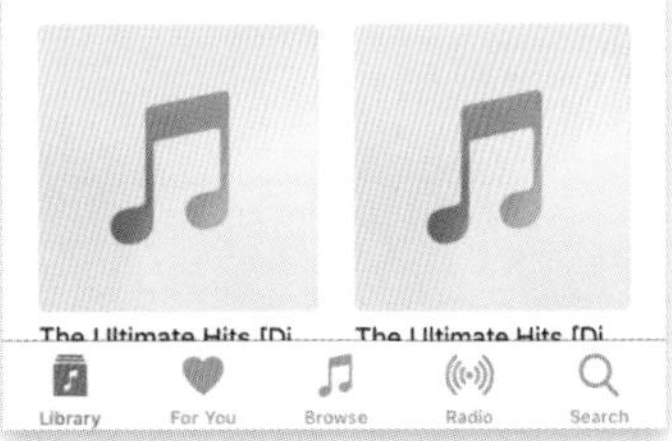

the ***Search*** tab is where you can search for songs, artists, albums, playlists, radio stations, TV shows, and movies in your own library or in Apple Music's catalog.

APPLE MUSIC'S PLAYBACK OPTIONS

When you play music in the Apple Music app, a player bar will appear at the bottom of your screen. Tap the bar to open the Now Playing screen, where you can play, pause, fast-forward, or rewind a song, and control the music's volume. Scroll down a bit and you'll also see two buttons labeled ***Shuffle*** and ***Repeat***. Tap ***Shuffle*** when you want the app to randomly play whatever playlist is selected. Tapping ***Repeat*** once will repeat the entire playlist once it has finished; tapping it twice will repeat a single song; and tapping it a third time will clear the repeat.

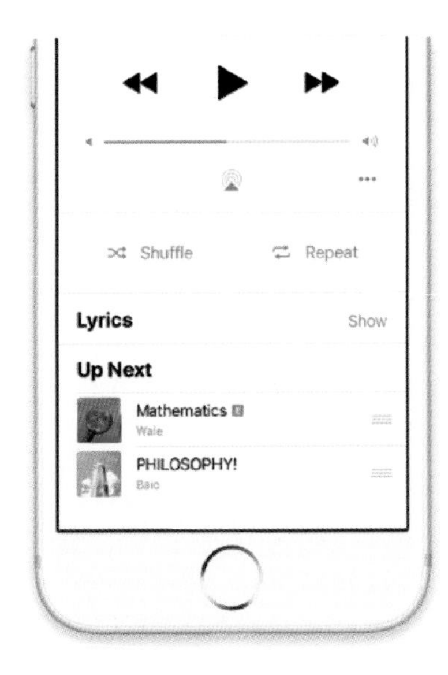

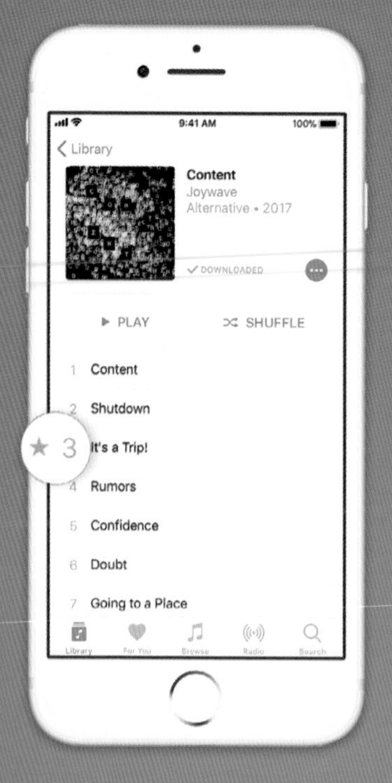

ADDING MUSIC IN APPLE MUSIC

When you want to add songs to your library, go to *For You*, *Browse*, or ***Search*** to see what's available in Apple Music. When you find a song or album you like, you can add them to your library by tapping the **PLUS SIGN (+)** to add a single song, or ***+Add*** to add the entire album. When searching for songs, you might notice a star next to some of the song titles. This symbol means that the song is one of the most popular songs on Apple Music.

WHAT IS APPLE'S HOMEPOD?

Brand new in 2018, Homepod will be Apple's version of a "smart speaker." Where the Echo has its "personal assistant," Alexa, Homepod will use Apple's tried and true voice assistant, Siri. The Homepod boasts seven tweeters in its base, plus a four-inch, upward facing woofer. The speakers are able to sense the configuration of a room, so the Homepod will be able to optimize the sound depending on where you place it in your home. All of this is said to create an immersive sound experience unlike any other smart speaker on the market.

The speaker also features six microphones to pick up voice commands. The microphones are sensitive enough that they can differentiate between your voice and any background noise, including the music played by the

speaker. So even if a song is playing, you'll be able to ask Siri questions like "who sings this song?" and still receive a response!

To be compatible with the speaker, you'll need an iPhone 5s or later running iOS 11. Homepod will be specifically set up to work with Apple Music, so you will need an Apple Music subscription to use it.

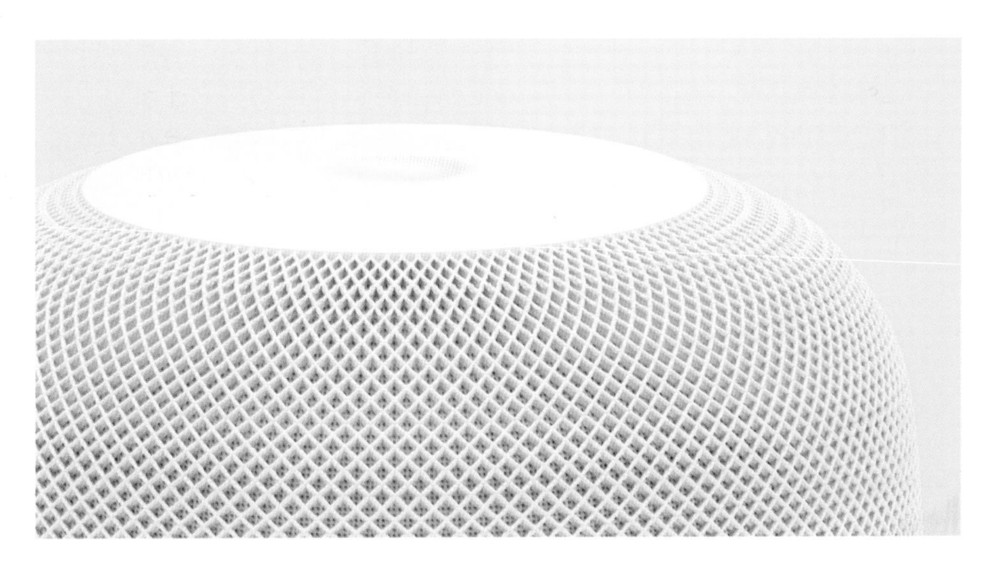

WHAT IS SOUNDCLOUD?

In 2007, Swedish artists Alexander Ljung and Eric Wahlforss came up with the idea of creating a platform where musicians could share their recordings with each other. The result was SoundCloud, which began as a way for mostly grassroots and aspiring musicians to swap music. But the app is now also a streaming service and distribution platform. Musicians love the service for sharing their songs with fans, and fans love it for keeping up with their favorite musicians.

SoundCloud also includes a social-network type model, where users can follow each other, comment on their favorite songs, and Repost songs in the same way a user on Twitter can Retweet a comment. Users can also join groups, where fans of the same organizations or genres can share and comment.

SIGNING UP FOR SOUNDCLOUD

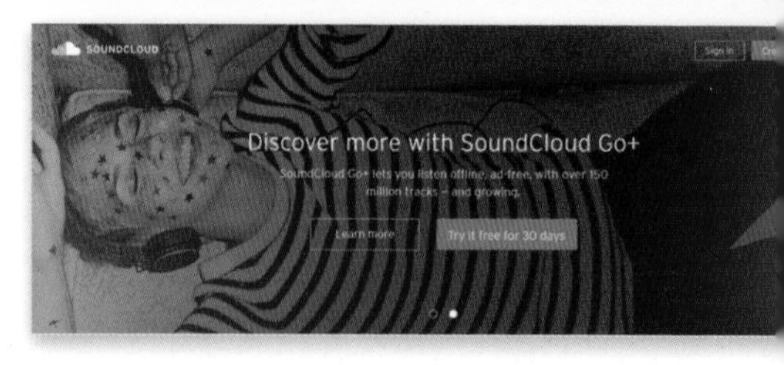

To sign up for SoundCloud, you can either use your email address, or link it to your Facebook or Google account and use the same information you use for those accounts. To get started, open a web browser and go to Soundcloud.com. Then, click the ***Create Account*** button. A window will pop up that asks you whether you'd like to continue with your Facebook or Google accounts, or you can enter your email address. If you use Facebook or Google, simply sign in using

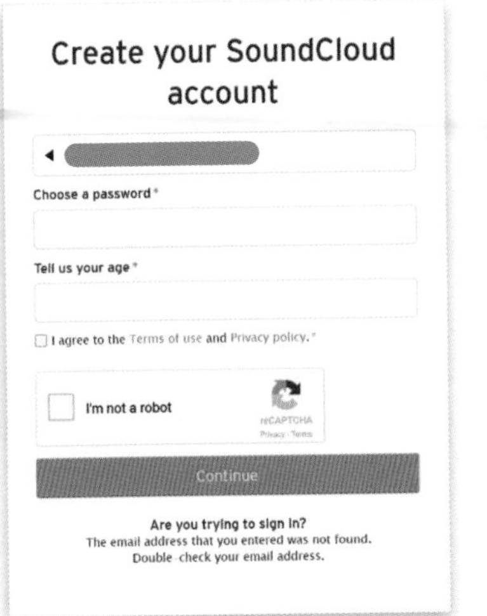

You will need an email address and a solid password that you will remember in order to create your account.

your usual information. If you want to use a different email address, enter your email address and then create a password for SoundCloud.

You'll need to check the box agreeing to the terms of use and privacy policy, and you may also need to check the ***I'm Not a Robot*** box to prove that you're human and not a computer. The final step is to create a username—this will be the name that appears on your profile. If you sign up with your email, you'll receive a message asking you to confirm your email—click on the link to confirm your email, and you're ready to use SoundCloud!

Signing up for SoundCloud is free, but the service also has two pay tiers: SoundCloud Go and Go+. Each one gives you access to millions of songs, with no ads. The first time you sign up for one of the pay services, you are given a free trial. So if you enjoy using SoundCloud, one of the pay options may work for you.

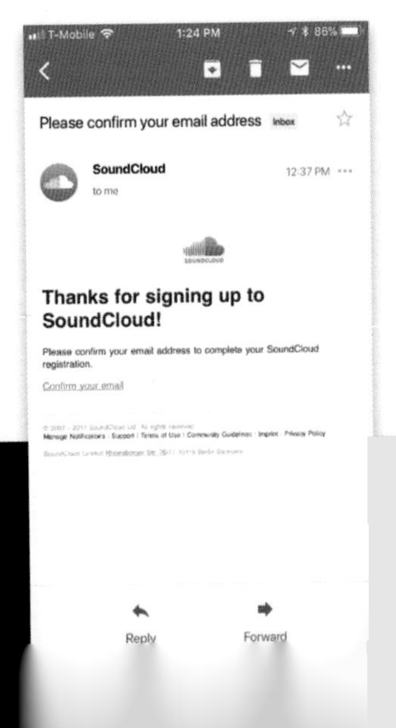

After you register, SoundCloud will send you an email to confirm your information. Follow the link in the email to finalize your registration.

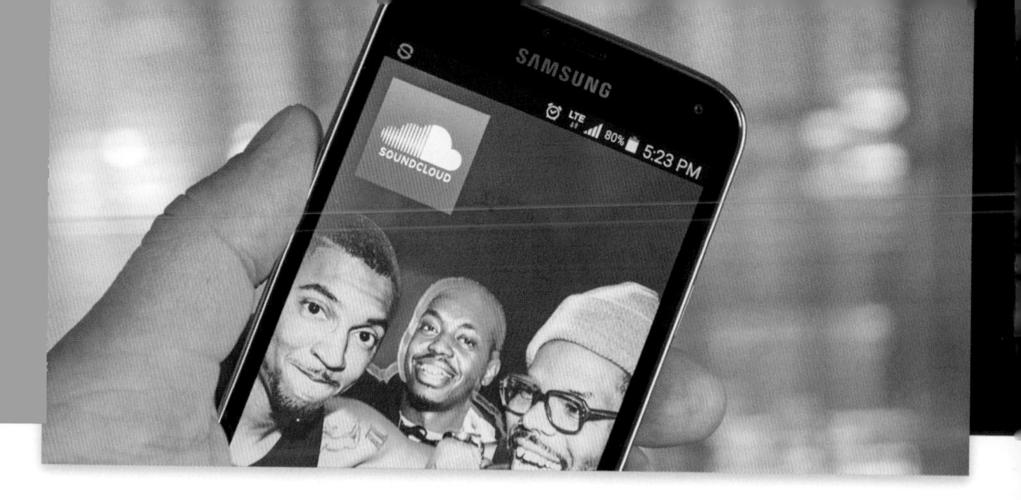

SEARCHING FOR MUSIC

There is a universal search bar at the top of the SoundCloud site, where you can type names of songs or artists you'd like to hear. SoundCloud provides you with a list of results, which can then be filtered according to what you're specifically looking for. If you're using the free service, the search result will also show you a list of tracks that are accessible with a SoundCloud Go subscription. This can help you decide whether paying for the extra features available in Go is worth the cost.

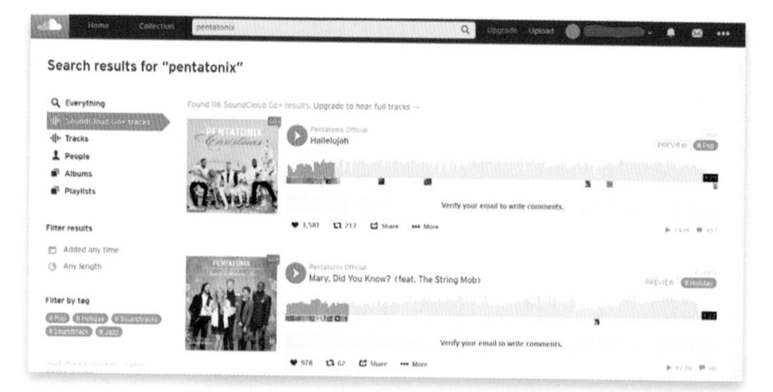

By conducting a universal search, you can search through people and artists, tracks, albums, playlists, or everything related to the search term. On the left of the screen you can begin to filter your results to help you find the track you are looking for more easily. Filters allow you to narrow down your search by locations, tags, when it was added, the track length, when the user joined SoundCloud, or by any assortment of other specifications. Note that your search term will provide results for track titles or artist names, but also playlist names, locations, or tags. The various tags a track is categorized under can be found in the gray box on the top-right of the track preceded by the **POUND SIGN (#)**.

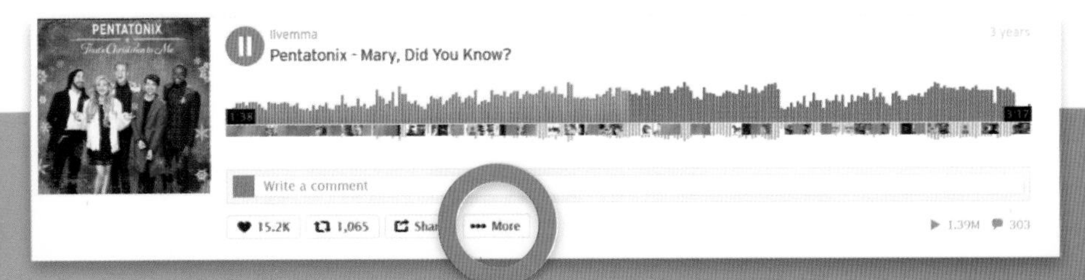

CREATING A PLAYLIST

To create a playlist on SoundCloud, click the ***More*** box underneath the track you like, and then select ***Add to Playlist***. You can then name the playlist, and choose whether it will be private or public. Once you've created a playlist, you can add more songs to it, or create a new playlist. SoundCloud recommends adding no more than 250 songs to a playlist, in order to make loading the songs easy and preventing delays.

Add to playlist Create a playlist
Filter playlists
Pentatonix
Add to playlist

Adding music to playlists and creating new playlists is easy. Once you have so many playlists that it becomes difficult to search through them, use the ***Filter Playlists*** field to narrow your search.

WHAT IS BANDCAMP?

Bandcamp's website proclaims that you can "discover amazing new music and directly support the artists who make it." This was the idea behind its inception in 2007. Artists—many of them independent newcomers—can upload their music to the site, where it can be played for free by site users. Tracks and albums may be purchased at varying prices, allowing users to pay more for songs they especially enjoy, helping to fund the musicians.

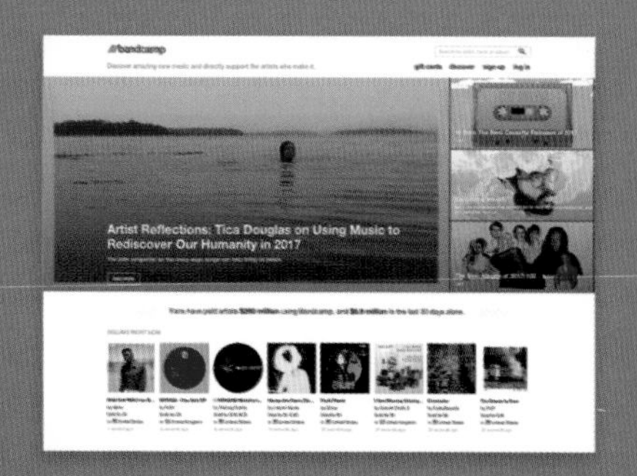

The artists who contribute to Bandcamp are given their own customizable "microsites" within the music service, where they can share their music and set their own prices for their songs. Users are given access to each artist's individual page, giving them a way to learn more about their favorite musicians and buy their music and merchandise. As of 2017, fans have paid more than $256 million to the artists on Bandcamp, making it a unique music company for both artists and fans.

Bandcamp is a great resource to help you discover all types of new music. Serving as a platform for both well-known and independent artists alike, Bandcamp provides a much needed service to record labels, musicians, and music fans.

As you can see, Bandcamp caters to all sides of the music industry, providing an economically viable way for music to be produced and consumed.

Sign up for a Bandcamp fan account

Email address

Password

Username

I have read and agree to the Terms of Use.

Please confirm that you're not a robot:

I'm not a robot

reCAPTCHA

Privacy - Terms

Sign up

Already have an account? Log in.

To sign up you will need an email address and a strong password that you will not forget. Create a unique Username that you will be identified as on the site.

Click on the *Sign up* button to activate your profile. You can also follow the links to download the Bandcamp app in order to listen to your music on your mobile device.

HOW TO USE BANDCAMP

To begin using Bandcamp, simply go to Bandcamp.com and click on the ***Sign Up*** button at the top right-hand side of the screen. A window will pop up asking which type of account you'd like: you can sign up as a fan, an artist, or a label. Most of us will fit into the fan category; when you click ***Sign Up as a Fan***, another window will pop up asking for your email address, and then you'll need to create a password and a username. As with SoundCloud, you'll be asked to agree to the terms of service and confirm that you're "not a robot." Bandcamp will then send you an email so you can confirm your account. Once you click the ***Activate Your Account*** link in the email, you're ready to go!

The general layout of an artist's microsite is pretty straightforward. You have the options of buying the release in the formats it is offered in, along with the track listing, and the option to preview the tracks. Some artists even offer deals, like we see here, to buy their entire discography for a bundled price.

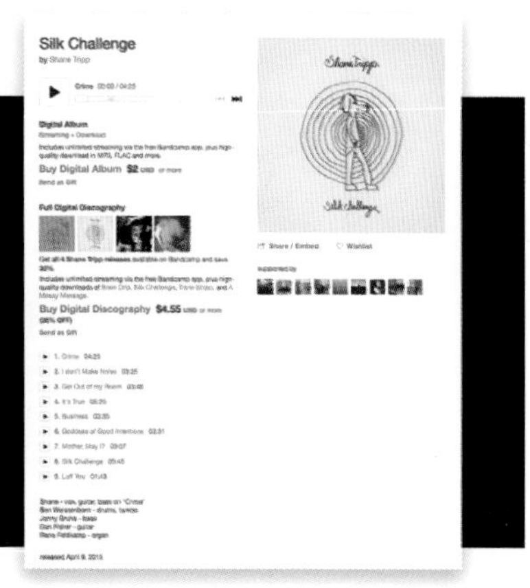

SEARCHING FOR MUSIC ON BANDCAMP

Once you log in to Bandcamp, you'll see a few buttons at the top of the screen, including ***Feed***, ***Collection***, and ***Discover***. Click ***Feed*** when you want to connect with other fans—you can follow other users with similar musical tastes, and see what they're listening to. You can also follow your favorite artists. ***Collection*** takes you to your own personal page, where you can add a picture and profile. This is where the songs you collect will be available. And when you click on ***Discover***, you are shown suggestions of new music, arranged by genre. Use the tabs at the top of the page to browse everything from hip-hop to country to classical.

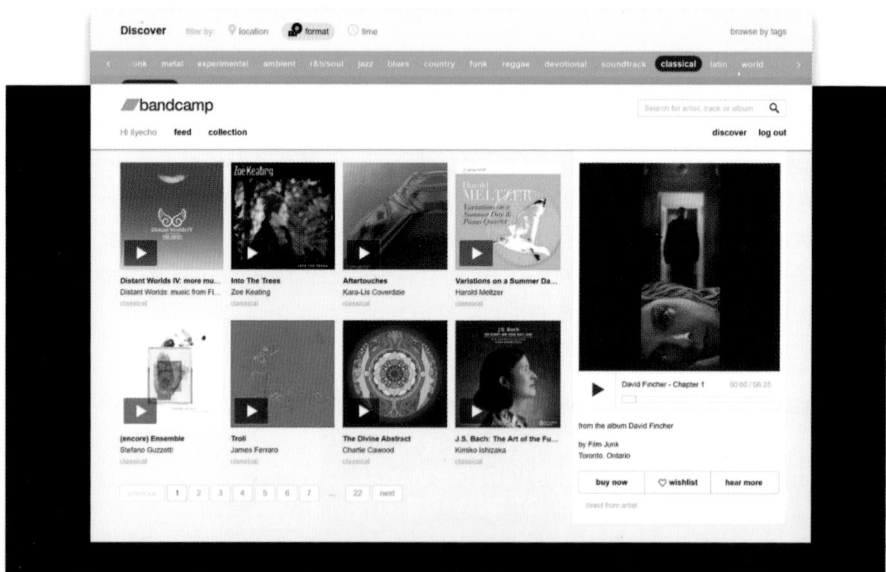

You can browse by genre in the green bar and then filter your results by location or format (digital, vinyl, cassette, or CD). The universal search will allow you to search by artist, track name, or album name.

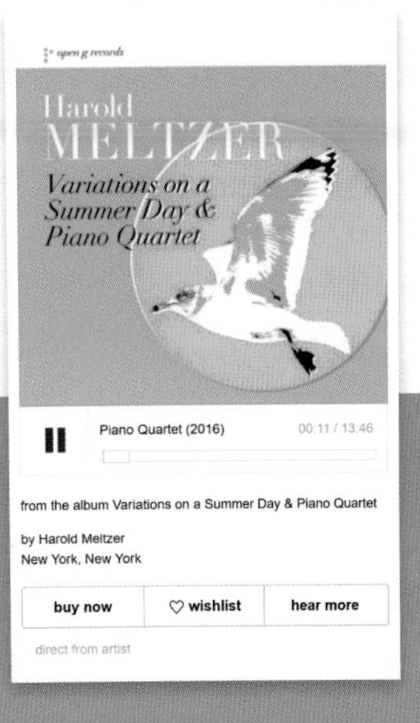

You can scroll through the song to preview it before you buy it or add it to your Wishlist to remember it for later.

LISTENING TO MUSIC ON BANDCAMP

You can listen to songs for free on the website, by clicking on the track you'd like to hear. The album cover of whatever you're listening to will appear on the right-hand side of the screen, with a few option buttons. You can add the album to your ***Wishlist***, or click ***Hear More*** to hear other songs on the album. You can listen to some tracks and releases for free on the Bandcamp website—depending on the artists' preferences—but if you find an album you'd like to download so you can listen to it offline, just hit the ***Buy Now*** button. A window will pop up that will either give the price of the album, or ask you to name your own price (usually with a minimum price listed). Then click either ***Add to Cart*** if you want to buy more, or ***Check Out Now***. You can either pay with PayPal or credit card.

After you check out, you'll be directed to a download page, where you can select a format for the download. The default format of MP3 is the best choice for home listeners. When you buy an album on Bandcamp, it will be in a compressed format called a ZIP file. You'll need to "unzip" it: if you're using a Mac, simply double-click the file. If you're using a PC, right-click the file and then choose ***Extract All***. That's it!

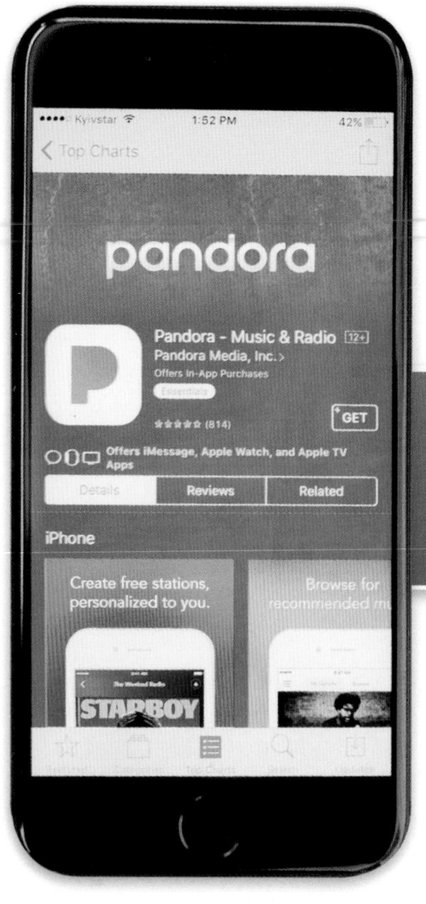

WHAT IS PANDORA?

Pandora is a music streaming service, which is also known as Pandora Internet Radio. Founders Will Glaser,

Jon Kraft, and Tim Westergren envisioned a service that would provide each user with an individualized "radio station" based on personal preferences. Users can respond to songs with the THUMBS UP or THUMBS DOWN icons, and Pandora uses the favorable and unfavorable ratings to create personalized stations. The service is free, although advertisements are

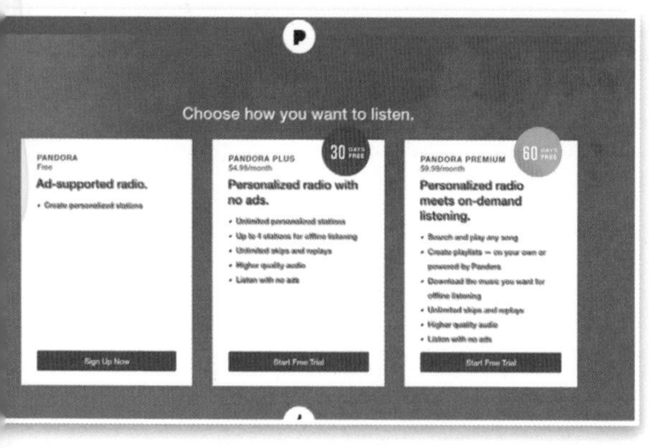

There are various levels you can sign up with. Pandora Plus and Pandora Premium have ad-free programming with different capabilities and features, while the free version contains ads and fewer capabilities.

played after every few songs. If you prefer a commercial-free listening experience, there are also pay subscription options.

SIGNING UP FOR PANDORA

Getting started with Pandora is easy: just go to Pandora.com and click the ***Sign Up*** button in the top left corner of the screen. You'll be asked for some basic info, including an email address and password. Pandora also asks for your birth year and zip code, in order to better personalize your music. It only takes a few seconds to sign up—after that comes the fun part: exploring the music!

CREATING STATIONS

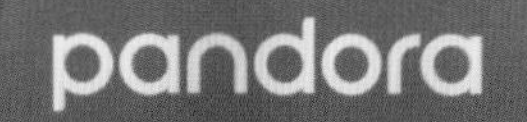

In the ***Create Station*** bar at the top of the screen, just type whichever artist, song, or genre you're looking for and hit ***Enter***. Pandora will begin playing something based on your search. If you like the song, you can hit the **THUMBS UP** icon at the bottom of the screen, and Pandora will find other songs that are similar to the one liked, and automatically play them next. If you don't like a song, hit the **THUMBS DOWN** icon, and the song will skip and move to the next song.

SEARCHING FOR MUSIC WITH PANDORA

When you search for artists and songs on Pandora, you automatically create a new station. If you'd like to search for music without creating a station, simply type your search in the ***Create Station*** box without hitting ***Enter/ Return***. A list of songs or artists will pop up. Click on ***See All Search Results*** at the bottom of the screen to see all of the available music you can listen to. If you don't see anything you like, click ***Clear*** on the right side of the ***Create Station*** box and try another search.

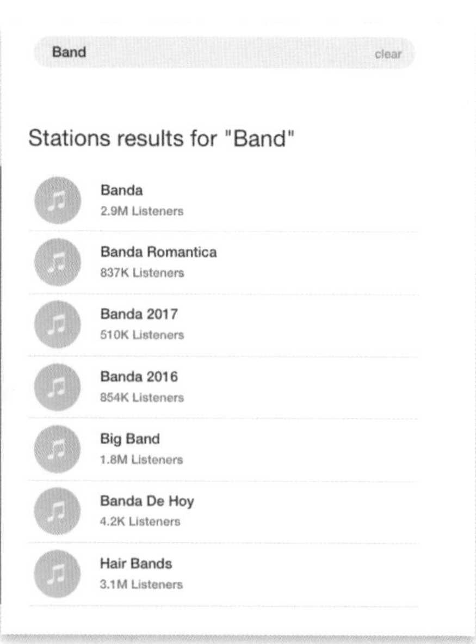

Typing in search terms without hitting the *Enter/ Return* button will bring up search results for different artists or stations that meet the criteria. By clicking *See All Search Results*, you can view the results by track, artist, or station.

Keep in mind that when you create a station based on an artist, Pandora won't necessarily play only music by that artist. The service chooses music that is similar to the original search, but could be performed by a different singer. For instance, a Frank Sinatra holiday station may also play holiday songs by Bing Crosby, Dean Martin, or similar artists.

Artist Audio Messages

On

Turn on to get updates from your favorite artists.

Station Created From

Jazz

4M Listeners

Thumb History

1

0

The Station Details window will give you options to edit and customize your radio station.

EDITING STATIONS ON PANDORA

If you like, you can edit or delete your Pandora stations. Click on ***My Stations*** in the top left corner of the screen, and then hover over whichever station you'd like to edit. Click on the ELLIPSIS icon at the bottom of the station's box. This will bring up several choices: ***Station Details***, ***Share***, and ***Delete***. ***Share*** allows you to share songs to Facebook or Twitter. And ***Delete*** will delete the station.

Station Details will bring up a new screen where you can edit certain aspects of your station. Clicking the PENCIL icon to the left of the PLAY button will allow you add a description of your station. Clicking on ***+Add Variety*** will bring up a list of different songs or artists that you can add to your station.

WHAT IS SPOTIFY?

Spotify is another music streaming service that gives users access to millions of songs, as well as podcasts and videos. Like Pandora, listening to music on Spotify is free, but advertisements run after every few songs. A Premium paid subscription is available that allows you to listen to unlimited songs without commercial interruption.

Spotify customizes its listening experience for each user by providing a Daily Mix of suggested songs as well as a Discovery Weekly playlist.

SIGNING UP FOR SPOTIFY

To create an account on Spotify, simply go to Spotify.com and click ***Sign Up***. You can either sign up using your Facebook info, or use your email address. After filling out some general information, hit the green ***Sign Up*** button and you're ready to go!

Once you've signed up, you can download the Spotify app straight to your desktop so your music will always be easy to access. Go to Spotify.com/download—the download should automatically start within a few seconds. Next, look for the app in your Downloads folder and double-click it, then follow the installation instructions. Now you can log in and listen to music.

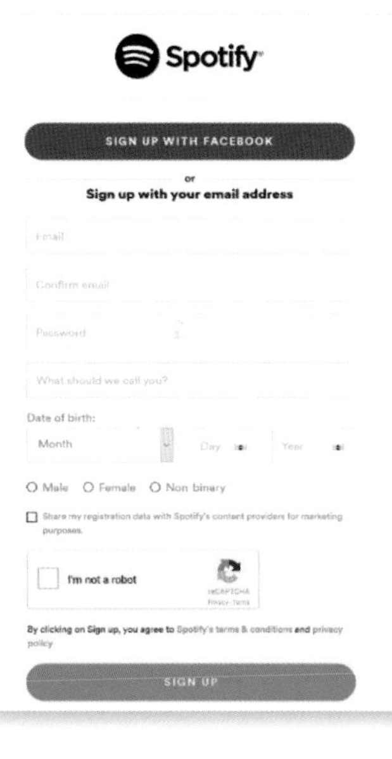

To sign up for Spotify, you will need an email address and a solid password that you can remember. You can also choose a username that will identify you on the site.

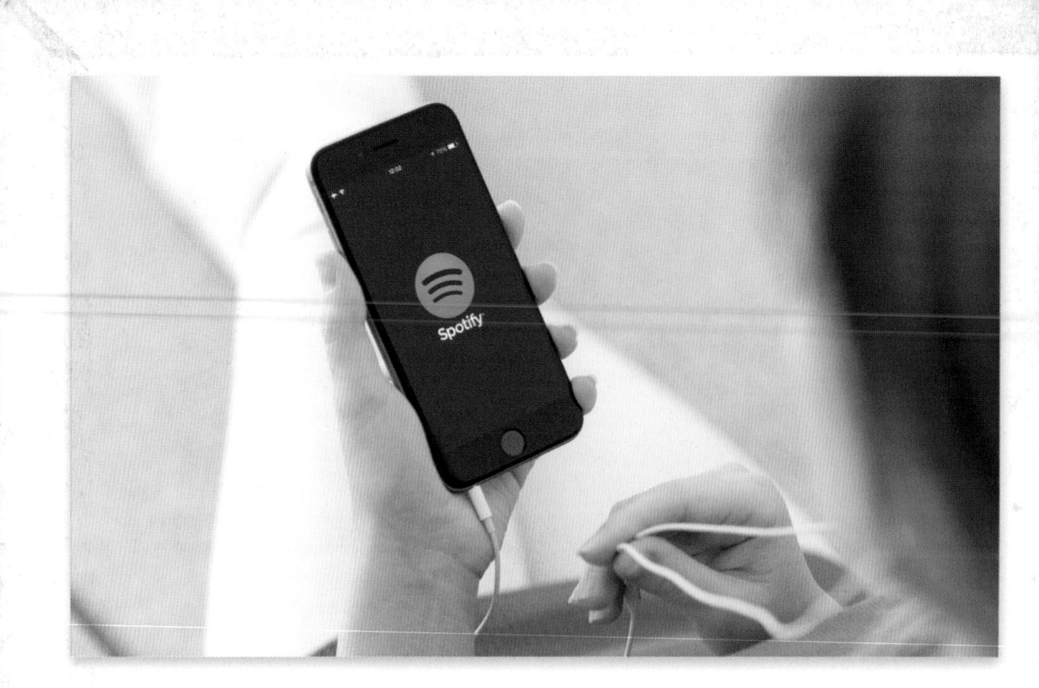

SEARCH FOR MUSIC ON SPOTIFY

To start listening to music on Spotify, type the name of an artist, song, or album into the ***Search*** field at the top of the screen. You can also refine your search by being more specific: for instance, typing “Frank Sinatra 1965.” This will provide you with a list of results from that year.

Play songs by using the **PLAY**, **PAUSE**, **PREVIOUS**, and **NEXT** icons at the bottom of the screen. You can also repeat songs or shuffle through all the songs on an album or in a playlist. When you find a song that you particularly like, you can click on the **PLUS SIGN (+)** to the left of the track title. This will add the song to your personal song library.

You can use the search bar at the top-left of the screen to find artists, songs, albums, playlists, podcasts, videos, genres, or other Spotify users.